THIS HEXAGONAL GRAPH PAPER NOTEBOOK BELONGS TO:

About Brickshub
Brickshub is a design company based in England. We specialize in custom apparel, personalized products, and branded merchandise. We understand that we're not just designing shirts, jewelry, footwear, watches, journals, tote bags, drinkware, or home accessories; we are helping people make memories. They're keepsakes that will remain with you for years to come.

Thank you for being our customer and for allowing us the opportunity to be of service.

Printed in the USA
CPSIA information can be obtained
at www.ICGtesting.com
LVHW082358251124
797634LV00038B/1341

9781728782959